Afro Hair: Procedures and Techniques

Also published by Stanley Thornes (Publishers) Ltd:

W. E. Arnould-Taylor & A. Harris	*Principles and Practice of Electrology*
J. F. Rounce	*Science for the Beauty Therapist*
E. Almond	*Safety in the Salon*
A. Gallant	*Beauty Guide 4: Epilation Treatment*
J. Gardiner	*Cardiac Arrest – What do you do?*
J. Gardiner	*The ECG – What does it tell?*
S. Henderson & M. Phillips	*Hairdressing and Science: A Competency-Based Approach*
W. G. Peberdy	*Sterilisation and Hygiene*

Afro Hair: Procedures and Techniques

Ginger Browne

Chairman and Founder, Caribbean Afro Society of Hairdressers, UK
Honorary Secretary, Hairdressing Training Board
Chairman, London area, National Hairdressers Federation

Stanley Thornes (Publishers) Ltd.

First published in 1989 by:
Stanley Thornes (Publishers) Ltd
Old Station Drive
Leckhampton
CHELTENHAM GL53 0DN
England

British Library Cataloguing in Publication Data
Browne, Ginger
    Afro hair: procedures and techniques.
    1. Hairdressing. Techniques
    I. Title
    646.7′242

    ISBN 0–85960–662–2

Typeset in 10/13pt Optima 55 by Tech-Set, Gateshead, Tyne & Wear
Printed and bound in Great Britain at The Bath Press, Avon

For my dear parents, Sugar Pet and Dollar Bill, who nurtured, encouraged, and inspired me throughout my life.

Contents

Ginger Browne

Preface

In order to obtain employment in hairdressing, one must be competent and capable of doing the job required. Recently, the Hairdressing Craft formed themselves into the Hairdressing Training Board and launched the National Preferred Scheme (NPS) in Hairdressing. Briefly defined it identifies clearly, precisely and concisely what hairdressers are supposed to *know* and what they must be able to *do* in order to perform the wide range of tasks involved as they cater for the needs of their clients.

Hairdressing does require a wide range of practical skills, but it also requires the application of a vast body of knowledge.

In the acquisition of a certificate in hairdressing, the assessment is on-going, i.e. it is carried out throughout the training. It involves oral questions, written work, and assessment of performance. It is validated by the City and Guilds London Institute. The NPS is divided into a series of 'units' or learning 'outputs'.

To obtain the Foundation Certificate in Hairdressing at present requires competency in certain specified 'units'. When the student has achieved this level of training, the way is clear for specialisation, i.e. colouring, Afro-Caribbean hairdressing, selling, etc. This book will be helpful to students who want to *begin* their specialist training in this very exciting and lucrative area of work.

Remember, hairdressers are known by their work; and their work is the result of the application of their knowledge and skill.

The author hopes that this book will assist you in your training and education.

A word of caution

This textbook is written for the student who has previously trained in European type hair, has completed a course in basic science, and who is at present enrolled on a course of instruction in Afro-Caribbean hairdressing, leading to a nationally recognised qualification, under the tuition of a teacher who has acquired an Afro-Caribbean Hairdressing Teacher's Certificate.

This book was never intended to be used as a tool in self-teaching or open learning. Therefore, the author accepts no responsibility for the misapplication of knowledge gained from reading this book without professional and experienced guidance.

Success in your career

Did you know that more than two-thirds of the world's population is non-white? Just think of the supply of potential clients! You may have heard the saying 'a good hairdresser never goes hungry'. You can verify this for yourself. Just look around and see.

What makes good hairdressers outstanding is the way in which they *apply* their knowledge and skill to satisfy the demands of their clients. This book has been written to assist in the acquisition of some of the basic knowledge that you will need to get started in building a successful career in hairdressing.

The teacher is important

There are certain portions of this textbook which require 'in depth' explanation. For example, the science sections need a trained and qualified teacher to illustrate their meaning and relevance in the context of hairdressing; for a practical demonstration of some of the methods and techniques, you need an experienced teacher to guide you safely through the application of caustic chemical creams on the hair. The length of time that the chemical can safely remain on the hair is of crucial importance, but *assessing* this length of time is a matter of judgement, depending on a variety of circumstances which are dealt with in the text, e.g. characteristics of the hair. Someone could be slow but accurate; and yet finish with disastrous results! 'To read and follow manufacturer's instructions' is not enough. Good judgement is also required. The teacher is there to help you develop that.

Ginger Browne
Bedford 1989

Acknowledgements

The author is deeply grateful to Robert Johns, Michael Coleman, Kevin Whirley and Stanley Thornes for their advice and encouragement in writing this book.

CHAPTER 1
The Structure of Hair

The structure of hair

All hair regardless of type, colour or ethnic origin grows from a region situated below the surface of the skin known as the hair follicle. At the base of every hair follicle is a small region of active tissue called the papilla where cells grow and multiply. With the formation of new cells from beneath, those above are pushed up the follicle where they undergo a change in shape and become hardened by the formation of keratin, a tough, horny protein composed of carbon, oxygen, hydrogen and sulphur. As a result of enzymic action the cells also lose their nuclei rendering them dead. Consequently what emerges from the surface of the skin is a keratinised dead structure which explains why no pain is experienced whenever the hair shaft is cut (see Figure 1.1).

Figure 1.1
A mature hair follicle

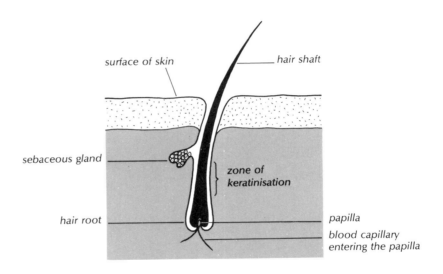

A single hair shaft consists of three distinct regions (see Figure 1.2). The outer layer, the cuticle, consists of overlapping, transparent scales arranged in a tightly-knit manner to keep water and other liquids out of the hair shaft. Directly beneath the cuticle lies the cortex, which forms the main bulk of the hair; it consists of bundles of elongated cells which lie parallel to the length of the hair. The cortex also contains pigment granules, which are deposited as it develops in the follicle. These granules are mainly responsible for providing colour to the hair. The innermost region of the hair shaft is known as the medulla and consists of a honeycomb (see Figure 1.2) of irregularly shaped areas of keratin interspersed with numerous air spaces. This region is not present in all scalp hairs, particularly if the hair is fine, and is not always found to run continuously along the length of the hair.

Figure 1.2
Longitudinal section
of a hair showing
the three regions

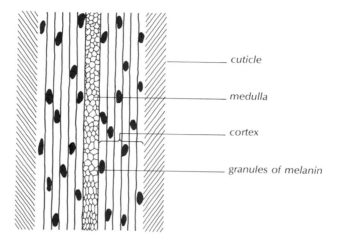

cuticle

medulla

cortex

granules of melanin

Various physical characteristics such as skin and eye colour, for example, are genetically determined, as is the type of hair an individual has. The characteristics of hair which are genetically determined range from colour, strength, texture, elasticity, length and degree of curl to baldness and premature greying. Although hair varies considerably from one individual to the next even within one family, it can on a general level be classified into three distinct racial groupings as shown in Table 1.1.

Table 1.1
Classification of hair
based on racial
groupings

Racial group	Colour	Pattern of growth	Hair shaft in cross section
Negroid	Dark	Curly	Flat
Mongoloid	Dark	Straight	Circular
Caucasoid	Light to Dark	Wavy/Straight	Oval

Afro hair

Genetics

There are various factors that disturb the genetic equilibrium of a population. Mutation and immigration increase the variety of a gene pool while emigration tends to decrease that variety. Immigration is the point of particular interest here, since through history it has played such a significant role in the lives of many people of African descent.

During the sixteenth century, it is believed that several million Africans, mainly from the Congo and Nigeria, but also from Angola and Senegal were taken to the Caribbean and North, Central and South America. It therefore followed, that with the introduction of new individuals to a given population an inevitable increase occured in the variety of possible characteristics that might be passed on to future generations. The tremendous amount of variation that could occur as a result of this is demonstrated by what an English anthropologist, Rowland Alexander, observed in 1853.

> If the African nations be examined, every possible gradation in the hair will be perceived, from the short close curls of the Kafir to the crisp but bushy locks of the Berberine, and again to the flowing hair of the Tuaryk or Tibbo. The Ashantees and others have hair which is rather curled than wooly and is occasionally so long as to reach the shoulders. The Foulahs or Fellatahs, natives of Sudan, have crepid and crisp, sometimes wooly hair. The Mandingoes in Senegambia have the genuine black, or frizzled, or wooly hair of the Negro. The flaxen locks of the Somali females (stained like those of the Lujean girls) render them conspicuous. The Calla tribes shave the head, preserving a lock of hair on it for every man they have killed. The Nubians have long, strongly frizzled or slightly crisp (but never wooly) hair. It is sometimes of a shining jet black, but in other cases of a color intermediate between the ebony black of Sennar Negroes and the brown of the Egyptians.
>
> *All about Health & Beauty for the Black Woman* by Naomi Sims, p. 54

In Jamaica today, 76 per cent of the population are of African origin, 15 per cent Afro-European, 1 per cent East Indian and the rest of Afro-Asian and Chinese descent. In Trinidad about half of the population are of African descent, one third of East Indian descent and the remainder mainly of mixed descent. (East Indian here refers to the descendants of immigrants from the Indian subcontinent.)

Afro hair

A unique structure

Afro hair is truly unique in both its structure and appearance. It is not really curly but more crinkly in appearance and very rarely does it grow very thick though the extreme frizziness suggests otherwise. The crinkly nature of Afro hair is believed to be due to the curved follicles producing over-short curls.

The keratinisation process is also believed to play some part in determining the unique nature of Afro hair. It is believed that keratin is distributed unevenly during the development and growth of the hair. It follows that if the hair shaft is uneven in strength it is more susceptible to breaking when subjected to pressure. Structurally speaking, therefore, Afro hair cannot withstand the same amount of stress as European hair.

As with hair in general it is possible to classify Afro hair, on a basic level, into four types: fine, normal, coarse and wiry.

- Fine hair is soft and spongy; in texture it feels like wool. This type of hair normally lacks the third inner region, the medulla.

- Normal or medium hair, as it is often known, lies between fine and coarse hair. Most people have hair which belongs to this category.

- Coarse hair may be described as 'hard' or 'crisp' to the touch and resistant and/or difficult to manage.

- Wiry hair is identified by its stiff, hard and glassy feel and appearance, and may be found in people of mixed race, i.e. Negroid and Caucasoid, and Negroid and Mongoloid.

Care and condition of Afro hair

The problem

Anything done to Afro hair can cause damage that leads to breakage. Even when hair is not chemically treated combs and 'picks' get caught in the crinkles and tear the cuticle. While combing the hair through, it will sometimes break off at the ends because of the uneven keratinisation process described earlier and the crinkly pattern of growth, both factors which result in the hair being structurally weakened. The ends may also be broken off while straightening the hair, relaxing it, or just trying to separate the split ends. When this happens a jagged split occurs. When that jagged split end touches other split ends near it the hair becomes matted. The process is continuous, almost 'contagious'. Hot combs and 'pressing' the hair make it easier to manage, quite simply by reducing the crinkles in the hair.

However when this process is applied every day the natural oils and proteins can be destroyed. This results in the hair losing its ability to retain its proper moisture balance, causing the hair to become dry and brittle, at which point breakage normally follows. The same problem may result from too frequent applications of relaxers and curly perms or wet-look.

The solution

- Reduce 'combing breakage' by using a detangling shampoo or conditioner that does not rely on resin and polymers in formulas.
- Avoid over-processing with chemicals that have a high alkalinity which tend to dry and weaken the hair.
- Avoid too frequent applications of strong chemicals.
- Be gentle when using combs, brushes etc.
- Condition the hair regularly with products which are specially formulated to:
  1. restore and regulate moisture balance
  2. strengthen the hair structure.

Regular trimming of the ends helps to strengthen the hair and improve its condition. Afro hair grows at a rate of approximately 1 cm ($\frac{1}{4}$–$\frac{1}{2}$ in.) per month (average), but because it may break frequently, it may appear not to grow. The ends should be trimmed once per month in order to reduce breakage from split ends. Hair that is maintained in this way will demonstrate better growth and condition.

The effects of chemicals on Afro hair

The reasons for straight or curly hair are not fully understood, though it is widely accepted that a straight follicle produces straight hair and a curved follicle curly hair. From this it may be inferred that the degree of curl a hair has is dependent on the curve of the follicle. As mentioned earlier, uneven keratinisation is also believed to play some part in the crinkly nature of Afro hair.

By chemically treating the hair it is possible to straighten Afro hair (relaxing) and also to produce a permanent wave. This occurs as a result of the chemicals producing a permanent change to the structure of keratin.

The two chemicals most commonly used to relax or straighten hair are: sodium hydroxide and ammonium thioglycolate.

Sodium hydroxide

During application, it has the effect of swelling and softening the main disulphide bonds which link together the polypeptide chains of the hair structure. The disulphide links lose one molecule of sulphur. After rinsing they rejoin as lanthionine links between the polypeptide chains and stop the chemical action. Hair which has been chemically treated in this way *does not return to its original structure* (see Figure 1.3).

Ammonium thioglycolate

For use as a hair relaxer it is usually available as a gel or thick cream. The effect of the thioglycolate on the hair is to break the disulphide links and soften the hair. Once this occurs the hair may be stretched with a comb or by hand to the desired straightness.

Neutralising

The neutraliser is left on the hair long enough to penetrate the cortex and stop the action of the thioglycolate. Thorough rinsing removes all of the relaxer and neutraliser from the hair and restores it to its original chemical state. It is essential to keep the hair straight during neutralising and rinsing if you are to achieve satisfactory results.

Like all proteins, keratin is made up of amino acid units. Each amino acid is composed of a different arrangement of the chemical elements carbon, hydrogen, oxygen and nitrogen which are chemically bonded together. The amino acids cystine and cystein also contain the element sulphur.

All amino acids have at least one amino group and one carboxyl group. The amino group of one amino acid can be joined chemically to the carboxyl group of another amino acid to produce a long chain of amino acids known as a polypeptide chain. In keratin the chains are arranged in a spiral manner called an alpha-helix (α-helix), like a coiled spring. Cross-linkages join up the chains into groups of three and also form between the coils of chains themselves.

The amino acid cystine can incorporate itself into two separate polypeptide chains since it has two amino groups and two carboxyl groups. This results in a bridge forming between the two chains known as a disulphide bond. It is this bond that is central to hair relaxing and producing a permanent wave.

In the case of relaxing the hair, a reducing agent which relaxes or breaks the disulphide bonds between the sulphur atoms of each molecule of cystine is applied to the hair. This results in two molecules of the amino acid cystein being formed. With the structures that enforce the crinkly nature of Afro hair removed it is possible to straighten the hair (see Figure 1.3). The reducing agent is then rinsed away and hydrogen peroxide is used to oxidise the cystein to re-form cystine which forms new disulphide bonds that maintain the new shape of the hair. In hairdressing this process is known as neutralising. To produce a permanent wave the process is almost identical, the only difference being that when the hair is ready to take on its new shape it is set in curlers or rods.

Figure 1.3

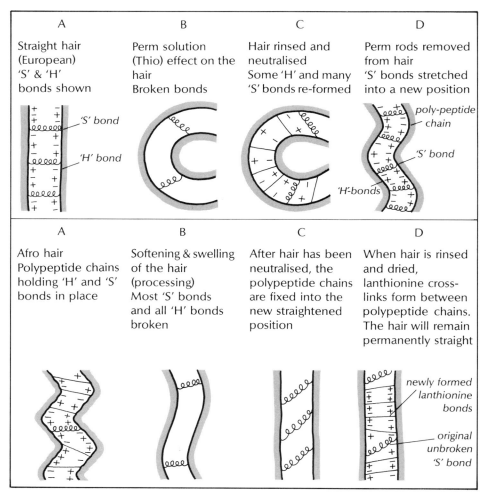

The degree or percentage of relaxation determines the new hairstyle. Relaxing the hair by 50 per cent will reduce the natural curl pattern by half. If the hair was very curly to begin with, a 50 per cent relaxation will give it an 'Afro look' that can be blow dried straight when desired. A 75 per cent relaxation of the hair will expand the natural wave pattern which may then be set on rollers or blow dried and tonged. Relaxing the hair by 85 per cent will produce a very weak curl pattern and hair may be styled in any manner desired. A 100 per cent relaxation produces straight hair with no visible curl pattern. At this level of relaxation structural damage occurs to the hair, which makes it undesirable unless it is the result that is sought in order to produce a particular hairstyle (see Figure 1.4).

Figure 1.4
Hair relaxing at
various processing
phases

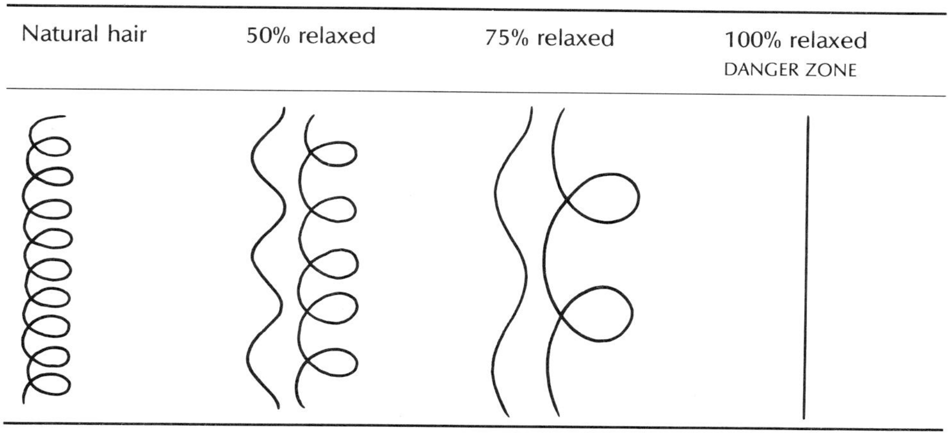

Natural hair	50% relaxed	75% relaxed	100% relaxed DANGER ZONE

Processing phases

Upon application of the relaxer there appears to be no change in the hair, but as the action of the chemical on the hair starts to take effect, i.e. swelling and softening the hair strands, the hair becomes 'rubbery' and more pliable. In order to monitor the processing the hair should be checked at regular intervals of 1–2 minutes.

Phase I   The outer layer of the cuticle is dissolved.
Phase II   The inner layer of the cuticle is dissolved and the strand of hair becomes smooth (see Figure 1.3, Afro hair A–D).
Phase III   The cortex layer is now exposed to the direct action of the relaxer. This is the critical time during the process and must be monitored very closely.

*Figure 1.5
Hair remains
straight when
processing is
complete*

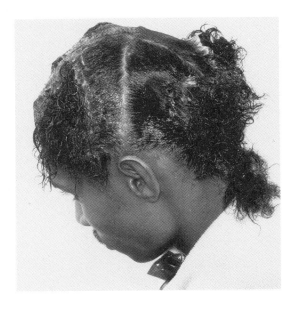

It is absolutely essential to follow the manufacturer's instructions very carefully regarding processing time. When monitoring, press the hair firmly against the client's head, gently stretch the hair downward in a straight line and release the pressure. If the hair remains straight, processing time is complete and the relaxer should immediately be rinsed from the hair using warm water.

*Figure 1.6
Chemical
straightener
brought
through
the length
and ends
of the hair*

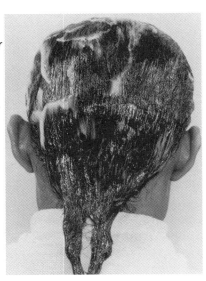

*Figure 1.7
Chemical
being
rinsed
from the
hair*

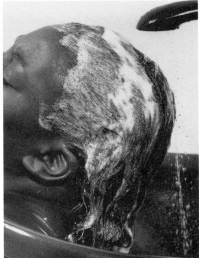

The technique used in producing a permanent wave on European (straight) hair is the same as producing a permanent wave on Afro hair. However before the procedure of winding the hair on perm rods, establishing a wave pattern, neutralising, etc. can begin, the natural curls of Afro hair must be removed by relaxing or softening (see Figure 1.8).

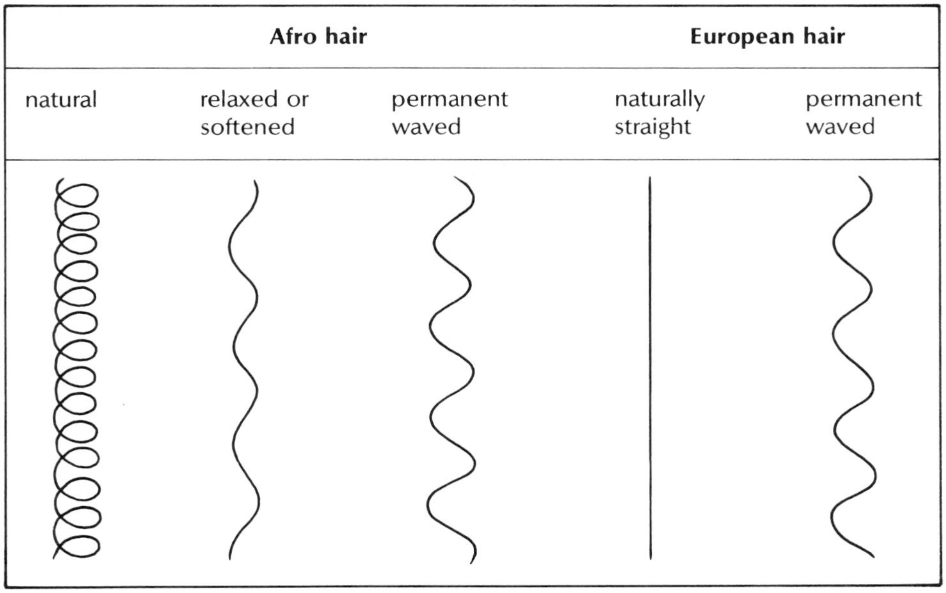

Figure 1.8 Producing a permanent wave on Afro and European hair

Afro hair			European hair	
natural	relaxed or softened	permanent waved	naturally straight	permanent waved

Sodium hydroxide (lye) in a solution with stearic and oleic acids was the first chemical manufactured for the use of straightening hair. The straighteners vary in their content of sodium hydroxide from two per cent to seven or more, and in pH value from 7 to 14. The more sodium hydroxide, the higher the pH value, and the quicker the reaction of the chemical on the hair. The chances of the quicker-reacting chemicals damaging the skin or hair are also greater.

The client's skin should be carefully protected with a 'base'. Some relaxers contain chemicals which can burn the skin and scalp, and if left in contact with the hair too long, may result in hair loss. Simply stated, hair relaxers change the crinkly nature of Afro hair into a straight form and leave the hair feeling soft, silky and smooth and easy to manipulate into fashionable and attractive styles.

The products used in hair relaxing

There are many relaxers available today to the hairdresser, their formulas varying according to the active ingredient which they contain. The most commonly available ones are:

- Sodium Hydroxide (lye)
- Calcium Hydroxide
- Potassium Hydroxide
- Ammonium Thioglycolate

The neutraliser for the first three usually comes in the form of a shampoo and for the last in the form of a watery liquid. The most widely used neutralising elements are hydrogen peroxide or sodium bromate. The neutraliser releases oxygen into the disulphide bonds, stabilising them in their new position. When neutralising with sodium bromate or hydrogen peroxide, neither should be left on the hair for too long. *Follow the manufacturer's instructions.* If the hair is not neutralised it will return to its naturally curly state.

Moisturisers and curl activators are used to groom and maintain the 'curly perm'/wet-look. These are the names that are given to the technique of producing a permanent wave in Afro hair. When they are not used the hair is dry and brittle-looking. However, following application the curls reappear. Some curl activators and curl moisturisers contain a high percentage of glycerine in their formulas. When too much glycerine is used on the hair it becomes coated and sometimes it is very difficult to penetrate. This could present problems when applying other chemical treatments to the hair, i.e. colouring, relaxing, perming.

Lye relaxers are those products that are manufactured with sodium and/or potassium hydroxide. Anything other than these two compounds is referred to as no-lye. Recently, a relaxer has been developed through combining calcium hydroxide and guanidine carbonate to produce guanidine hydroxide. This formula has proved to be less irritating to the skin than some lye-based relaxers while at the same time being just as effective.

Thioglycolate perms, commonly referred to as curly perms or wet-look perms, are not compatible with hydroxide-type (lye and no-lye) relaxers. The reason

*Figure 1.9
Man with curly
perm; woman with
relaxed and
highlighted hair*

for this is that the curl process uses strong oxidising agents as neutralisers, namely hydrogen peroxide and/or bromide salts.* These salts tend to weaken the hair. The no-lye hydroxide relaxers are not compatible with curly-perms etc, mainly because of the oxidising agents used in the process.

Hair and scalp treatments are a means of proper maintenance of the hair. Hair receives nourishment from the blood such as amino acids which are derived from the capillaries of the papilla and brought to developing hair cells by tissue fluid. Certain scalp treatments are designed to increase the flow of blood to the papilla. Proper massaging also helps in directing and increasing the flow of blood and nourishment towards the papilla. (See Figure 1.1.)

Various scalp treatments are available in order to correct or maintain different situations. Electric currents (such as the Telsa current), steams and violet-ray,

*Salts are compounds that are formed by the reaction of acids and bases, with water also produced by the reaction. Two common salts and their formulae are sodium chloride (table salt), which contains sodium and chloride, and magnesium sulphate (Epsom salts) which contains magnesium, sulphur, hydrogen and oxygen.

along with electric and manual vibrators, are all forms of stimulating the scalp to promote the growth of hair. Hair and scalp treatments must always begin with a thorough analysis of both the hair and the scalp to determine which treatment will best benefit the individual.

Health and safety advice

Special care should be taken when treating a client with high blood pressure, a heart condition or a nervous disorder. Discussion with the client about any medication which is currently being taken, whether internal or external, is absolutely essential. Never perform an electrical treatment on a client with a pacemaker.

Diet, nervousness, fatigue or lack of general care are all factors that affect the hair and scalp. Through consultation determine what hair care products, if any, are being used and how often they are being used. These are all factors that should be considered when planning a course of hair and scalp conditioning treatments.

Proper maintenance of the hair is sometimes a problem for some people. By using a pH-balanced shampoo and conditioner the proper pH balance of the hair and scalp is maintained.

The majority of chemical hair straighteners, other than thio types, contain sodium hydroxide. Regardless of any variations in names and formulations the only difference is in the percentage of caustic they contain. The following precautions should always be observed:

1. The client's hair and scalp must be in good condition.
2. Do not brush or shampoo the hair *before* the application of the relaxer as damage to the scalp may occur causing irritation when the relaxer is applied.
3. Use a protective base cream around the ears, hairline and on the scalp.
4. The stylist should wear rubber gloves.
5. Prevent the relaxer from coming into contact with unprotected skin, eyes and ears.
6. Avoid over-processing.
7. Rinse thoroughly with warm water at a backwash-type basin.

Most 'no-base' relaxers (except thios) are caustic. The base is usually in the emulsion, and the hydroxide is in a lower concentration.

Hydrogen peroxide decomposes readily into water and oxygen. The reaction can be caused by heat or light and is the reason why hydrogen peroxide, in its unstable state, should be stored in a cool dark place and purchased in containers or brown bottles which do not admit light. If it does decompose in the salon, it simply turns into water and cannot perform its function. The function of hydrogen peroxide is to release oxygen in the bleaching and colouring of hair, and to stop the action of cold wave lotions, i.e. ammonium thioglycolate.

Acids, bases and salts

Many of the substances used in hairdressing fall into one of these categories of chemical compounds.

An acid is a compound which contains hydrogen atoms that are positively charged. These are called hydrogen ions and give acid its properties. An acid has a sour taste and turns litmus paper red.

A base, also called an alkali, is a combination of a metal with a hydroxide ion. A base has a bitter taste, is corrosive to the hair and skin, soapy to the touch and turns litmus paper blue.

A salt is a compound formed by the union of an acid and a base. The hydrogen of the acid is replaced by a metal, and the other product formed is always water.

The pH scale

The relative acidity of a solution is commonly given in terms of 'hydrogen ion concentration' calculated on a scale from 0 to 14 called the pH scale, which means potential hydrogen ions. Pure water (distilled) has an equal number of hydrogen ions (H^+) and hydroxide ions (OH^-) and has a pH of 7. As a result this is considered the neutral point. Solutions that have a greater hydrogen ion concentration than the hydroxide ion concentration are acidic. Those that have a greater number of hydroxide ions than hydrogen ions are alkaline or basic (see Figure 1.10).

It is very important to know the pH of the various products used in the salon. If a product is strongly acidic, i.e. with a pH of 4 or lower, or strongly

Figure 1.10
The pH scale

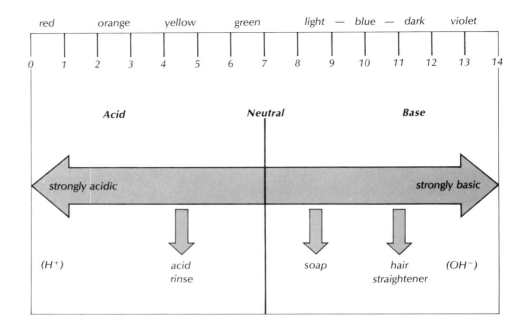

alkaline with a pH higher than 10, it may be harmful to the skin or hair. The pH value should be obtained from the manufacturer of any product with which you are not familiar. It can also be determined by a colorimetric indicator, a few drops of which are added to a small amount of the solution. The pH value is then read by colour comparison with a chart. A far more accurate way of measuring the pH of a solution is through the use of a pH meter, which makes use of the fact that the voltage of an electric current passing through a solution will change according to the pH of the solution. The average pH values of some common products are:

Product	Approximate pH
ammonium thioglycolate (active ingredient in curly perm/wet-look)	9
boric acid	5
soap	8–9
soapless shampoo	6–7
sodium bicarbonate (baking soda)	8–9
1% sodium hydroxide (used in relaxer)	12

CHAPTER 2

Working Practices

Working practices

Amongst the many services that the modern hairdresser can offer the client there are certain procedures and techniques that form a general prerequisite to the various services and treatments available. They may be necessary to protect the client's clothing and ensure comfort, or for the success of a particular treatment; or they may be compulsory in the interests of health and safety before a certain treatment can be given.

For obvious reasons it is essential that time and patience are spent perfecting such techniques and ensuring that they are performed correctly. Time invested in learning these general procedures and techniques will pay off when working in the salon, where a personal interest in the clients' welfare and appearance will obviously contribute to the reputation of a stylist.

Gowning the client

Before beginning any form of treatment on the hair it is important to gown the clients to protect them and their clothing from chemicals, water spills and splashes.

Figure 2.1   Three stages involved in gowning a client

Begin by requesting the client to be seated; this is done to ensure that the client is comfortable and relaxed and also to allow the stylist, particularly one who is short, to operate easily, comfortably and, above all, efficiently. The next stage involves the removal of all jewellery, including glasses, which could be damaged or broken during the service and obstruct the work of the stylist. If there is a collar on the client's clothing, fold it down inside, place a towel around the shoulders and draw a gown over them, fastening it at the back of the neck. The process of gowning is completed by placing a towel over the shoulders (see Figure 2.1). There are now three layers of protection on the client, which should be more than sufficient for a careful stylist.

Shampoo

The purpose of a shampoo is to cleanse the hair by removing grease (sebum), dirt, dead skin scales and the remains of various hairdressing preparations. It should also leave the hair in as near to its normal condition as possible, feeling soft, smooth and manageable.

Most shampoos used in the salon are surface-active detergents which include soaps and soapless (non-soap) types. The action of detergents depends on the presence of hydrophilic (water-loving) and hydrophobic (water-hating) groups of atoms. These combine to form molecules that possess both the above properties, having a hydrophobic tail and a hydrophilic head (see Figure 2.2).

Figure 2.2
The structure of a
detergent molecule

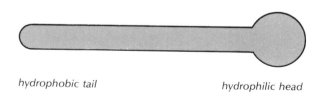

hydrophobic tail   *hydrophilic head*

On its own, water is not a good cleansing agent because its molecules have a tendency to stick together forming droplets; furthermore it fails to penetrate oil. This is due to the strong forces of attraction between the molecules. At the surface these forces produce an inward pull which makes the water act as though its surface has a skin. This is called surface tension. When detergent is added to water, the hydrophilic heads push in between the water molecules at the surface thus expanding it (see Figure 2.3). This is known as lowering the surface tension of the water causing it to 'wet' the surface with which it is in contact.

*Figure 2.3
The action of detergents on the surface of water*

The hydrophilic heads cause the surface of the water to expand.

hydrophilic heads

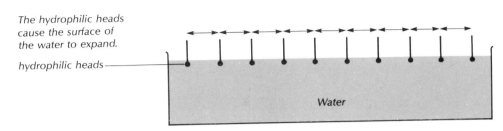

When oil and water are shaken together, tiny droplets of oil become suspended in the water forming an emulsion. Eventually the two liquids separate into two distinct layers. To maintain a permanent emulsion the addition of an emulsifying agent is required; detergent acts as such an agent. Thus in the process of removing grease the hydrophilic heads of shampoo molecules, which also carry a negative charge, enter the water while the hydrophobic tails enter the grease. The negative charges on the hydrophilic heads repel each other causing the grease to form tiny droplets that can be easily dislodged from the hair. This process can be further helped through the use of warm water and the action of rubbing the hair (see Figure 2.4).

*Figure 2.4
The cleansing action of detergent*

Hydrophobic tails enter the grease.

Negative charges cause the grease to roll up.

Droplets of grease form an oil-in-water emulsion.

Water

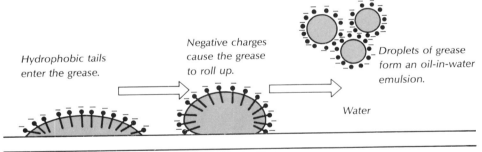

strand of hair

The process of shampooing not only cleanses the hair but also prepares it for any subsequent treatment. Through the action of detergent, water can enter the cortex more easily enabling the fibres to stretch, thus assisting in the setting processes. There are various additives which may be added to a shampoo, such as colours, fragrances, conditioning agents, preservatives, thickeners and germicides. However the various factors which distinguish a good quality shampoo are as follows:

1. The shampoo should be safe to use on the hair and scalp and not be too acid or alkaline.
2. It should spread easily and also evenly over the hair.
3. It should produce a rich, smooth lather that is produced quickly and remains stable.
4. The emulsifier should not be so strong that the hair and scalp are left so de-greased that dermatitis and dry unmanageable hair result.
5. The shampoo should not allow the grease droplets to join up into larger globules and be deposited back on the hair.
6. It should be easy and quick to rinse from the hair.
7. The hair should be left in a clean and manageable condition that can be combed easily.

Shampooing

The stylist must first ensure that the hair is free of all adornments. The client can then be positioned at the shampoo basin, preferably a backwash basin. Correct positioning is very important both for the comfort of the client and also for correct shampooing. It is best to have the client sitting upright or as close to this position as possible. It is absolutely essential to test the temperature of the water on the inside of the wrist. To prevent scalding, always turn on the cold water tap first at a moderate rate of flow and then slowly introduce the hot water. When a comfortable temperature has been reached proceed to wet the client's hair. At this point it is important to ask if the water temperature is comfortable since sensitivity varies from one individual to the next. Adjust as necessary and proceed to wet the entire head of hair.

Apply a measure of shampoo, distribute it evenly over the hair and massage gently. If a particular treatment is to follow, it is best to massage the hair only and not the scalp. This is to avoid the risk of making the scalp too sensitive, or causing abrasions.

Generally shampoo is applied to the hair twice. With the first application little lather may be produced because of the application of grease to the hair. When this is rinsed away the second application of shampoo should produce more lather.

Having completed the shampooing process, rinse the hair thoroughly removing all the shampoo; a good indication that this has been achieved is when the water runs clear. Use the hands gently to remove excess water from the hair and bring the client back to an upright position, whilst at the same time arranging the towels in such a manner that the wet hair is contained and the client's clothing is kept dry. The way in which this is done successfully requires practice and is really a matter of personal technique.

Drying the hair

Towel drying should be performed gently to prevent the hair from being damaged or further tangled. Care should also be taken not to remove too much moisture from the hair as it prevents damage from occurring during the setting process; as a guideline the hair should feel barely damp.

Blot the water from the hair rather than using a rubbing motion to dry it. Do not let water from the scalp run down the face and into the eyes and ears of the client, as it is uncomfortable, annoying and reflects badly on the stylist. Having completed the towel drying process the hair should be combed using a large-toothed comb.

Safety tests

When applying any chemical preparations to the hair it is essential to read and follow the manufacturer's instructions carefully. Deviation from these instructions should not be made without a thorough understanding and working knowledge of what can be accomplished on various textures of hair.

Before any specialised chemical treatment can be performed it is vital to ensure that the client is compatible with the chemical to be used and will not suffer any adverse reaction. Most responsible manufacturers supply details of the tests that should be performed before proceeding with a specific treatment. However, a good stylist should have a knowledge of the basic tests normally associated with treatments such as hair relaxing and colouring.

Hair relaxing should not be performed on hair which has recently been straightened by a hot pressing comb. When specifically requested to do so, hair which has been damaged by 'pressing' should be given a series of conditioning treatments to restore some of its elasticity before the relaxer chemicals are applied. The amount of treatment given depends upon the condition and texture of the hair.

For obvious reasons a record card should be kept of all clients who have had their hair chemically treated. The information recorded should include the client's name, address, texture of hair, treatment received, date of treatment, name of stylist, product used and result. Other information that may be relevant, such as pregnancy, nervous condition, occupation, etc., should also be recorded.

Examining the scalp

This is best done using the forefinger to part the hair, starting at the right temple and working clockwise around the back of the head at intervals of approximately 1 cm ($\frac{1}{4}$–$\frac{1}{2}$ in.) (see Figure 2.5). The parted section should be held between the forefinger and middle fingers with the palm of the hand facing the hairdresser. The unparted hair should be held between the fingers of the left hand. It is best to make straight partings.

The handle of a styling comb may also be used to part the hair. However, care must be taken to avoid scratching the scalp as a serious problem may be created when chemicals are used. Relaxer, for example, is very difficult to wash out without causing aggravation to the scratched area and a serious infection may also result. For the same reason relaxer should not be applied to hair where the scalp has a sore, scratch or abrasion.

Figure 2.5
The technique of
examining the scalp

The patch test

This is carried out prior to the application of a chemical such as relaxer to determine whether or not the client is allergic to it. There are two sites where it may be performed, either on the inside of the arm or behind the ear.

1. *Inside of the arm*   Using a piece of cotton wool wash the inside of the elbow (arm at the bend) with some mild soap. Dry gently by patting with a towel, do not rub. Apply a sample of the relaxer to be used over an area of approximately 4 cm ($1\frac{1}{2}$ in.) in diameter. This should remain on the arm for the length of time recommended by the manufacturer. The client is instructed not to brush or scratch the scalp and asked to return to the salon 24 hours later. If a red spot appears where the chemical was applied, the test is positive, indicating that the client is allergic. Consequently that particular chemical should not be used.

*Figure 2.6
Inside of the arm
patch test*

2. *Behind the ear*   An area behind the ear of approximately 4 cm ($1\frac{1}{2}$ in.) in diameter is prepared and treated in the same way as described in 1. Again, if a red spot appears, treatment should not proceed with the chemical used in the test.

*Figure 2.7
Behind the ear
patch test*

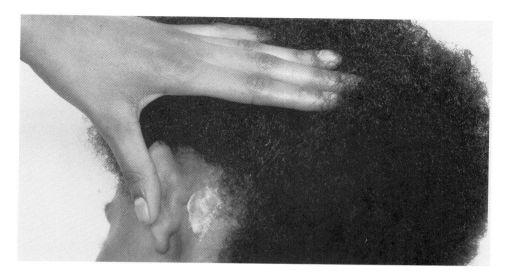

In either of the above cases, if the client complains of nausea or headaches even if no red spot appears, treatment should not proceed with the chemical used in the test.

Hair condition analysis

Having found chemicals compatible with the client, it is important to test the hair for elasticity and porosity, or to determine whether or not the relaxer can be successfully applied to the hair. This may be determined through the following tests.

1. *Finger Test* A single strand is selected from the crown area and 'run' between the thumb and forefinger. If the strand seems to 'stick' or feels 'bumpy', the hair has good porosity and would be able to absorb the chemical.

2. *Tension Test done on dry hair* This is designed to establish whether or not the hair will stretch, without breaking, under normal pressure. Normally, Afro hair has the ability to stretch about one-fifth of its natural length without breaking. Approximately six strands are selected from the crown area and held at the base with one hand, close to the scalp. With the other hand the strands are pulled. If the hair appears to stretch, or 'give'

under pressure, it has sufficient elasticity to take the relaxer. Failing this the hair could be suitably conditioned before application of the relaxer.

3. *Strand test* A small hole about 4 cm ($1\frac{1}{2}$ in.) in diameter is cut into greaseproof paper or aluminium foil. Approximately 8–12 strands are selected from the crown area and passed through the hole; relaxer is then applied direct to the strands and allowed to remain for 2–3 minutes after which it is removed with cotton wool. If the hair shows signs of damage, then this indicates that the application of relaxer should not proceed.

Hair colouring: skin test

This is also known as a predisposition test and should be performed 24 hours before each application to discover whether or not a person is allergic to the contents of the product. Allergy is an unpredictable condition and some clients may react to a product after using it for many years without a reaction. The skin test may be carried out on either the elbow (inside of the arm at the bend) or behind the ear extending to the hairline; the procedure is described below.

1. Select the test area which should be approximately 2.5 sq cm (1 sq in.) and wash with soap and water; dry gently by patting with a soft tissue, do not rub (see Figures 2.6 and 2.7).

2. Prepare the test solution consisting of one capful of the actual tinting mixture. If a mixture of colours is to be used, the test solution should be prepared to duplicate the actual tinting mixture. Mix one capful of peroxide using the same strength as would be used when preparing the solution for the actual service (see manufacturer's instructions).

3. Using a cotton wool-tipped applicator, apply enough solution to cover the prepared test area. Allow to dry, and leave uncovered, untouched and unwashed for 24 hours. Clothing should be loose to allow free circulation of air.

4. After 24 hours, if the test area shows no sign of inflammation or irritation, application of the hair colour may proceed. However, if the test area shows signs of inflammation or blisters, or the client complains of burning or itching, the client is allergic and the product must not be used.

Symptoms of hair tint poisoning are as follows:

- swelling
- itchy red spots/blotches all over the body
- tiny blisters containing a clear liquid
- headaches and vomiting.

A client showing any of the above symptoms must seek medical attention immediately.

Test for metallic dyes

Mix 30 ml (1 oz.) of 20 volume peroxide and 20 drops of 28% ammonia water (.880 ammonium hydroxide) in a glass or plastic container. Cut several strands of hair from the client's head, attach them to a strip of sticky tape and immerse those strands of hair in the above solution. Remove after 30 minutes and examine carefully.

Test for lead Hair will change colour at once, often becoming much lighter immediately.

Test for silver No reaction whatsoever as a peroxide and ammonia solution cannot lighten the hair strand because it cannot penetrate the silver coating, while the hair strand with no artificial colouring or penetrating tint will lighten to some degree.

Test for copper The solution will start to boil within a few minutes and the hair strands will feel hot and give off an unpleasant odour. After a few more minutes the hair is weakened and can be easily broken.

If any of the above results are obtained, the metallic coating must be removed before the hair can be successfully coloured, relaxed, curly permed or have any type of chemical treatment. Specific products are available for the removal of metallic and compound dyes; they should be used as described in the manufacturer's instructions.

Precautions

1. Carry out a skin test 24 hours prior to the application of the tint.
2. Do not apply tint if a skin test is positive.
3. Do not apply tint if metallic or compound dye is present in the hair.
4. Do not apply tint if the scalp is scratched, bruised, cut, sore, etc.
5. Use clean applicator bottles, bowls, brushes, combs, towels and any other tools required.
6. Do not brush the hair prior to a tint.
7. Do not apply a tint without reading and understanding the manufacturer's instructions.
8. Carry out a strand test for colour correctness (accuracy), breakage, if any, and/or hair discolouration.
9. Choose a shade of tint that harmonises with and complements the client's complexion and skin undertones.
10. Use applicator bottle or a glass or plastic bowl to mix the tint.
11. Do not mix the tint with the peroxide before it is ready to be used.
12. Protect the client's clothing by gowning correctly.
13. Always wear protective gloves during the application.
14. Do not use the mixture on eyelashes or eyebrows.
15. Do not overlap while doing a tint re-growth.
16. Throw away any unused tint mixture.
17. Use a low pH shampoo as an alkaline one may strip the colour from the hair.
18. Advise client about conditioning treatments, type of shampoo and conditioner to use at home between salon visits.
19. Allow at least a week before colouring, perming, relaxing, etc.
20. Complete a client record card, carefully noting the results.
21. Always wash hands before and after serving a client.

CHAPTER 3

Hair Straightening

Hair straightening

Afro hair may effectively be straightened either chemically or through the use of heat. Chemical treatment produces a more or less permanent result, whereas the effects of heat treatment are mainly temporary. The degree to which the hair may be straightened depends on how curly the hair is to begin with and the strength and type of treatment employed. As the hair grows it regains its natural shape, unless of course the new growth is also subjected to the same straightening process.

The product range

There is on the market a variety of hair relaxers specifically prepared to reduce the amount of curliness in Afro-type hair. They are produced in varying 'strengths' to accommodate the range of hair texture and conditions which exist, and to achieve a variety of effects. The most commonly used are those based on sodium hydroxide.

In the following discussion the relaxers will be identified by their active ingredients rather than by brand name in order to avoid the appearance of preference or prejudice. The relaxers divide into two main categories, which can be sub-divided in order to distinguish their similarities and differences. In the first category most of the relaxers are based on *sodium hydroxide*. This is available in two forms of application, *base* and *no-base*, and is produced, generally, in three strengths: mild, regular and super.

The *base* type requires a pre-application of protective petroleum (oil based) preparation whenever it is used.

The *no-base* type may be used without the pre-application of a petroleum preparation.

Within the first category are two other relaxers which are often referred to as 'no-lye' relaxers. However this is considered to be a misnomer. Respectively, they use *potassium hydroxide* and *calcium hydroxide* as their active ingredient (see page 12).

Within the second category is *ammonium thioglycolate* which was discussed earlier on pages 7 and 12.

Hair relaxing

This is the name given to the chemical process of straightening Afro hair. Various preparations are available, differing mainly according to the active ingredient present. They usually come in two forms, liquid and cream, the latter being more popular as it is much easier to control during application. Three strengths are available, mild, regular and super, the type used depending upon the texture of the client's hair. It is worth mentioning here that a single head of hair may include several textures. Aside from texture the two general characteristics which need to be considered are:

1. *Elasticity*   The ability of hair to stretch and return to its normal length without breaking under normal conditions (healthy hair can be safely stretched by about one-fifth of its length).
2. *Porosity*   The ability of hair to absorb moisture regardless of its texture.

Essential equipment required for hair relaxing

1. mirror
2. steriliser
3. hair cream (grease)
4. setting/styling lotion
5. neutralising shampoo
6. relaxer
7. conditioner
8. base/protective cream
9. rubber gloves
10. brush
11. combs
12. timer
13. cotton wool
14. neck strip
15. roller setting pins
16. pin curl clips
17. towels
18. rollers
19. hair net

Sectioning the hair

After a thorough examination of the scalp, the hair is gently brushed to the back of the head. The recommended way of applying the base and relaxer is to divide the hair into five distinct sections, the reason being that should the relaxer begin to work too fast on any particular section, it can be washed out quite easily without interfering with any other section.

To section the hair, a parting is first made from the centre of the left eyebrow to the crown, best done by using the tail of a styling comb. The same is done on the right side of the head, drawing a parting from the right eyebrow to the crown. Another parting is made, beginning from the top of the left ear to the crown; the top of the right ear to the crown; and the middle of the nape of the neck to the crown. It is also acceptable and satisfactory to divide the hair into four sections (see Figure 3.1).

Figure 3.1
Two patterns of
sectioning the hair

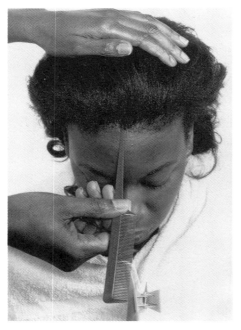
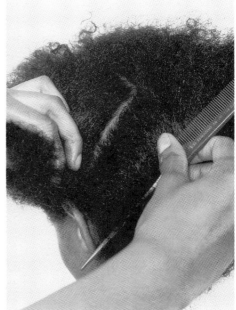
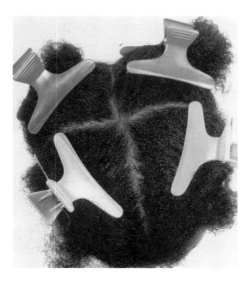

The application process

Most relaxers do not require the application of a base. Consult the manufacturer's instructions supplied with the relaxer to determine whether or not one should be used. However, where one is required it should be applied in the following manner.

Using the fingers, part the hair about 1 cm ($\frac{1}{4}$–$\frac{1}{2}$ in.) apart and apply the base freely by hand, taking care not to rub it on. Begin application at the nape of the neck and work upwards towards the crown, working around the entire head. Having done this, carefully examine the scalp to ensure that it has been thoroughly and completely covered. Failure to do this can result in a burn to the scalp by the chemicals present in the relaxer. Finally, the base is applied to the hairline.

In many cases when performing a new-growth application technique, it is necessary to apply a conditioner to the hair before the relaxer. Hair which has been hot combed, pressed, bleached or tinted is damaged and weakened and likely to break at the ends. Conditioner is applied to strengthen the damaged and weakened hair and protect it against further damage. It should be applied to the hair in the same systematic manner as the base is applied to the scalp.

Relaxer is also applied in the same manner, starting at the nape of the neck and working upwards and around the entire head. Extreme care should be taken not to let the relaxer spill onto the ears, scalp or face. Relaxers can be applied with fingers, comb or a tint brush. The stylist should wear protective gloves, whichever technique is used, throughout the application stage. Great care should be taken not to scratch the scalp.

The spreading phase

This is also known as the smoothing or stretching phase. The relaxer is spread over the hair for the following reasons:

- To be sure that the entire head is completely covered from the root area to the hair tip. (Applies to virgin hair relaxer application.)

- To be sure that the hair is completely relaxed from the root area.
- To be able to determine how fast, and where, the hair is being relaxed.

If a timer is recommended it is at this stage that it should be introduced. Spread the relaxer in the same way as the base, conditioner and relaxer were applied, starting at the section where the relaxer was first applied. The application of the relaxer should be closely monitored to determine how fast the hair is being relaxed. If the action of the relaxer is too fast in any particular section, it should be washed out immediately (see page 10).

Having worked around the entire head, spreading the relaxer over the root area, apply the relaxer to the hairline. Speed is essential at this stage with only a few minutes remaining before the relaxer is washed out; however, safety and a methodical approach should not be forgotten.

With the spreading process satisfactorily completed, the relaxer can be rinsed from the hair. A backwash basin should be used to avoid damage to the client's eyes and skin. Test the water to ensure that it is not too hot, as it will not only burn the client but also cause the hair to revert to its natural curly shape. If the water is too cold, it will not stop the action of the relaxer, and will not thoroughly rinse out all the relaxer and the base. Ideally, the water should be very warm but not hot.

Gently rub the shampoo into the hair and scalp, working from the front hairline down to the crown, taking in the sides. Care should be taken to keep the hair completely straight and to prevent the ends from becoming tangled. At this point the hair is much more fragile and tangled ends can be broken easily during combing.

The hair should be closely inspected by making small partings to ensure that all the relaxer and shampoo have been removed. If all the relaxer is not thoroughly washed out, it can result in breakage and possible hair loss. Some relaxers contain agents strong enough to actually eat away the hair. Manufacturers today usually incorporate the neutraliser in their recommended shampoos, the purpose of the neutraliser being to fix the hair in its new relaxed shape (see Figure 3.2). Finally, the hair should be conditioned. The hair may now be trimmed, cut and styled.

Figure 3.2
Procedures
following the
removal of the
relaxer

 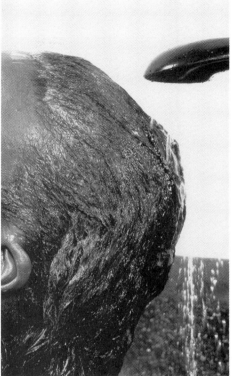

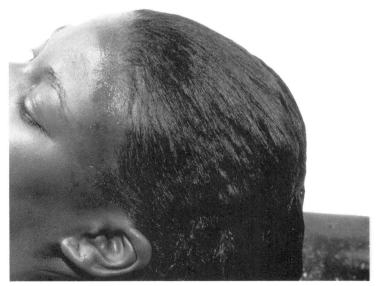

Thermal hair straightening

Thermal hair straightening, or hair pressing as it is often called, temporarily straightens Afro hair. It is not harmful to the hair, when done correctly, and prepares it for thermal curling and subsequent styling procedures. There are two basic techniques of hair pressing:

1. soft press, obtained by using a pressing comb
2. hard press, obtained by using thermal or 'hot' irons after using the pressing comb, or by pressing the hair twice.

Essential equipment required

1. pressing comb
2. stove, heater
3. pressing oil or cream
4. hairbrush and comb
5. shampoo
6. towels and cape
7. curling or thermal irons

Examine the hair and scalp and prepare the hair in the following way:

1. Shampoo, rinse and towel dry.
2. Apply pressing lotion, oil or cream.
3. Dry hair thoroughly.
4. Comb and divide the hair into four sections and secure.
5. Heat comb.

Procedure

Sub-divide the hair in each of the sections into 3 cm ($1-1\frac{1}{2}$ in.) partings, which is the recommended size for medium textured hair of average density. For hair that is thin or coarse, the following are recommended:

- thin or fine hair with sparse density – use larger sections
- coarse hair with greater density – use smaller sections in order to ensure complete heat penetration.

Pressing may begin at any point on the head; the important thing is to maintain continuity to ensure all the hair is adequately pressed.

1. If necessary, apply pressing lotion, oil or cream evenly and sparingly over the small sections of hair.
2. Test the temperature of the heated pressing comb on a piece of tissue paper. If the paper shows signs of scorching, let the pressing comb cool before proceeding.

3. Lift the end of a small section of hair using the index finger and thumb, upward and away from the scalp.
4. Take the pressing comb and insert the teeth into the top side of the hair section.
5. Draw the pressing comb along the hair for a short distance only and then quickly turn the comb around so that the hair partly wraps itself around the comb. The back of the comb actually does the pressing (see Figure 3.3).
6. Press the comb slowly through the hair until the ends of the hair pass through the teeth of the comb.
7. To achieve a hard press, press the hair twice on top, reverse the comb and press the hair once on the bottom side.

Figure 3.3
The technique of
hair pressing

A hard press is recommended when the results of a soft comb press are not satisfactory. The entire pressing procedure must be repeated. When pressing fine or woolly hair, the stylist should follow the same procedure as for normal hair, being careful not to use too much pressure. Hair breakage can be avoided by exerting less pressure on the hair near the ends.

In the case of coarse hair, greater pressure is required to ensure that the hair is adequately straightened. When pressing short hair, care must be taken at the hairline. Care must also be taken when pressing tinted, lightened or grey hair. In such cases, a moderately heated pressing comb should be used with light pressure. Avoid excessive heat on grey, tinted or lightened hair, as it might discolour it.

In all cases of hair pressing the process should be completed by applying a little pressing lotion, oil or cream to the hair near the scalp and brushing it through the hair. This may be followed by thermal, roller or other curling processes.

Figure 3.4
Hair that has been curled using thermal technique

Problem solving

Hair relaxing requires a great deal of expertise, skill, and product and hair knowledge, much of which can only be acquired through experience. Table 3.1 shows some of the more common mistakes made and ways of correcting them.

Table 3.1 Common problems involved in the hair relaxing process

Problem	Cause	Possible solution
Hair still natural looking.	Hair under-processed. Relaxer was removed from the hair too soon.	Correct the strength of the relaxer and match it to corresponding hair type/texture. Relaxer should be allowed to remain on the hair long enough to achieve the desired result.
Hair appeared straight before rinsing but returned to its natural shape after rinsing.	Hair was not sufficiently stretched during processing. Water temperature was too warm during rinsing/ shampooing.	Correct stretching technique. Rinse with warm water.
Client experiences irritation during processing.	Scalp may have cuts, abrasions or scratches. Hair was shampooed prior to relaxer application. Client may be sensitive to the active ingredient in the relaxer.	Do not apply relaxer if the scalp has broken skin or abrasions. Hair should not be shampooed at least 3 days prior to application. Find a suitable relaxer by doing a skin test.
Hair disintegrates during processing; breaks during the stretching process.	The hair has previously been treated with a metallic dye. Relaxer used too strong on colour-treated or bleached hair. Hair is very porous. Hair has been previously damaged.	Do not apply relaxer to hair treated with a metallic dye. Choose the relaxer strength best suited to hair type and condition. Do not apply relaxer to damaged hair.

Problem	Cause	Possible solution
Hair breakage occurs within one week of relaxing and continues at an abnormally high rate.	Sodium hydroxide type relaxer was used on hair previously treated with 'thio' chemical. Sodium hydroxide relaxer was not thoroughly rinsed from the hair.	Do not apply sodium hydroxide relaxer over a 'thio' relaxer. Do not use a strong relaxer on fine woolly hair. Do not apply relaxer to hair which has undergone prolonged pressing.
Root area straight while lengths and ends appear crinkly.	Application technique was incorrect. The virgin hair was treated with the relaxer from the roots to the end.	On virgin hair relaxer should be applied to the lengths and ends first, and to the root areas when processing has begun.
Hair shaft clearly shows area that has been relaxed while the new growth retains its natural shape.	When subsequent re-touches have been carried out, the relaxer has not been combed through properly.	When performing a re-touch service to straighten the new growth, the relaxer should be carefully applied to the new growth. Avoid double processing particularly during the stretching process.
Hair is broken and thin around the hairline, the remaining hair is very straight.	Hair has been over-processed. Hair was over-stretched and elasticity destroyed during processing. Relaxer strength may have been too strong for the type/texture of hair.	Correct strength of relaxer and match to corresponding hair type/texture. Correct stretching technique. Avoid leaving the relaxer on the hair too long.

CHAPTER 4
Curly Perm

Curly perm

The curly perm and wet-look are interchangeable terms used to describe the process of producing a permanent wave in Afro hair. There are two stages to the process, the first of which involves straightening the hair and the second which involves fashioning the hair in curls and fixing it in that state. There are two ways in which this may be done, either through the use of heat which produces a temporary set, or through the use of chemicals where the structure of keratin is changed resulting in a permanent set. Consequently, it is this process to which the term curly perm/wet-look applies.

The relaxer used to straighten the hair should only be the one supplied or recommended by the manufacturer of the curling agent which may itself be available in three strengths, mild, regular or super, depending upon the manufacturer. When selecting the preparations to be used on a client's hair, the same factors have to be considered as when selecting a relaxer:

1. texture or textures of the client's hair (Figure 4.1)
2. elasticity (Figure 4.2)
3. porosity (Figure 4.3).

Figure 4.1

Figure 4.2

Figure 4.3

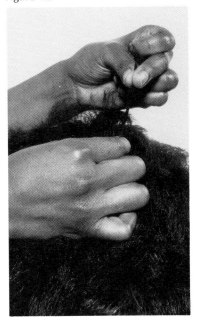

The process should not be attempted on hair that has been relaxed using products which contain sodium hydroxide (lye) or calcium hydroxide (no-lye) as the result obtained will be less than satisfactory. Hair which has been damaged by hot combing should be given a series of conditioning treatments to restore some of its elasticity. Having observed the conditions stated it is a worthwhile precaution to perform a test curl to determine whether or not the process should be performed.

Procedure for a test curl

A test curl should be performed on virgin hair as well as previously curled hair; this will reveal essential information such as the condition of the hair, likely processing time and whether or not the process will be successful.

Virgin hair should be straightened as directed by the manufacturer's instructions using the recommended relaxer. The hair should then be shampooed and towel dried. Select a thin section of hair from the front, back and side of the head and wind onto a curler each. Treat each curl with the chemicals that have been selected as suitable and then cover the hair with a plastic cap and process according to the manufacturer's instructions. After processing unwind each curler approximately $1\frac{1}{2}$–2 turns and check for 'S' formation; if the desired curl pattern is achieved, the process may proceed.

Application

Figure 4.4
Sectioning the hair

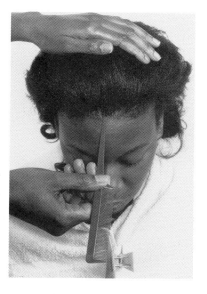

Part the hair from the centre of the forehead through to the nape of the neck and then from the top of the right ear across to the top of the left ear, dividing the hair into four equal sections.

Using the handle of a tint brush, separate the hair horizontally, at the nape of the neck, into 1 cm ($\frac{1}{4}$–$\frac{1}{2}$ in.) sections, depending upon the thickness of the hair. If treating new growth only, apply the chemical product (cream or gel) from the root outwards to the line of demarcation or end of new growth line.

The tint brush should be used in a similar manner as a paint brush, alternating it from side to side to wipe the cream or gel off the brush and onto the hair. At this stage it is good practice to ask clients if they are experiencing any irritation. Generally this should not be the case, since allergy to the chemical used should be determined before beginning the process by doing a patch test. If irritation is experienced, the chemical should be rinsed from the hair immediately. Otherwise application may proceed to the rest of the hair (see Figure 4.5).

*Figure 4.5
Application
technique for new
growth*

Beginning at the nape of the neck, part the hair into 1 cm ($\frac{1}{4}$–$\frac{1}{2}$ in.) sections again and comb the chemical evenly through the hair using a large tooth comb. It is essential to ensure that each and every strand is covered thoroughly in order to achieve a professional result. Continue to work around the entire head, following the same pattern as for application. Special attention should be paid to the hairline at the nape of the neck and behind the ears (see Figure 4.6). Following satisfactory completion of this process, cover the head with a plastic cap (see Figure 4.7).

*Figure 4.6
First section
completed*

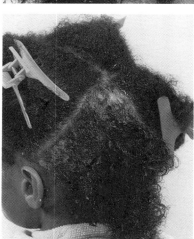

Figure 4.7

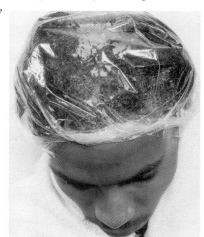

Processing

During processing the hair becomes soft and pliable; with the hair in this state, extra care must be taken not to damage it. Processing can be achieved either by body heat or through the use of a hood hair-dryer, which is faster than the former method. In order to determine when the hair has reached the correct stage of processing, remove the plastic cap and, using the back of a wide tooth comb, press the hair firmly and draw the comb down. If the hair remains straight when the comb has been removed, processing is complete. However if the hair returns to its natural shape, processing is not complete. This test should be performed at the crown of the head, back and sides.

Having determined that processing is complete, the hair may be wound onto curlers, the size of which can be determined when doing the test curl (see Figure 4.8). Note that in the first of the two pictures the hair being held by the stylist is unevenly distributed, the second rod from the front of the head is incorrectly placed and the third rod from the top, on the right side of the model's head, is incorrectly fastened. The second picture shows that these faults, which are very common ones, are being corrected as the stylist uses both hands. The tail comb is held while hair is wound onto the perm rods.

Figure 4.8
Winding hair onto
perm rods

Processing of the curls is done in the same way as for straightening the hair, covering the hair with a plastic cap and using the heat of the body or a hood hair-dryer. The processing time should be monitored closely and adhered to strictly as directed by the manufacturer's instructions.

To check the development of the curls unwind a curler in the area where the hair was first wound, by approximately $1\frac{1}{2}$–2 turns, and check for development of the 'S' wave pattern (see Figure 4.9). Again, perform this test at the crown, back and sides of the head. The various stages of processing are shown in Figures 4.5–4.8.

Figure 4.9
The 'S' wave pattern

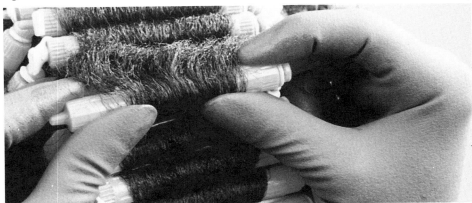

Figure 4.10
Rinsing

Once the hair has been sufficiently processed, it should be rinsed thoroughly with warm water. Where processing has been achieved by using a hood hair-dryer a period of 3–5 minutes should be allowed before rinsing.

Neutralising

The amount of neutraliser used should be as directed by the manufacturer's instructions. For the client's safety and comfort place a double strip of cotton wool around the entire head and apply the neutraliser to each individual curler. Start application at the crown of the head and work down and around the head until all the curlers have been soaked. Leave on for ten minutes. To test if neutralisation is complete, touch the head with the palm of the hand. If heat or warmth is felt in any area neutralising is not complete. Add another application of the neutraliser and check again in five minutes. When the head feels cool to the touch, the curlers may be removed carefully from the client's hair, disturbing the curls as little as possible. Neutralise the hair again ensuring that all the curls are soaked with the solution (see Figure 4.11). Wait for three to five minutes and rinse the neutraliser from the hair with warm water.

Figure 4.11
Neutralising the hair

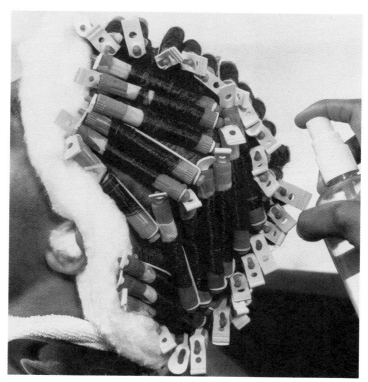

Remove excess water from the hair by towel drying and follow the manufacturer's instructions regarding the use of conditioner, moisturisers and curl activators.

Problem solving

Table 4.1 shows some common problems that may be experienced when carrying out the curly perm and offers some possible causes and solutions.

Table 4.1
Common problems
encountered in the
curly perm process

Problem	Cause	Possible solution
Hair disintegrates during processing; breaks when winding onto curlers.	The hair has been previously treated with a metallic dye. Too strong a preparation was applied to colour-treated or bleached hair.	Do not apply preparation to hair which has been treated with a metallic dye. Choose the preparation best suited to hair type, texture and condition.
Hair will not form into wave pattern after maximum processing time.	The hair has been previously treated with sodium hydroxide or similar relaxer.	Do not apply to hair that has been treated with sodium hydroxide or similar relaxer.
Hair breakage occurs within one week of the service and hair loss occurs at a very high rate.	The preparation was too strong for hair type, texture and condition. Hair was not thoroughly rinsed. Over processing.	Match correct strength of chemicals to appropriate hair type, texture and condition. Rinse all chemicals from hair thoroughly. Monitor development at short intervals of time.
Hair still looks natural.	Hair was under-processed. Manufacturer's instructions were not correctly followed. Curlers were too small for the length of hair.	Ensure that hair has softened. Follow the manufacturer's instructions carefully. Select the correct size of curlers for hair length.
Hair appears to be dry with no apparent curl.	Hair has not been treated with 'after care' products, i.e. curl activators, moisturising conditioning spray.	Thoroughly wet the hair. If the curl re-appears then the process was successful and the application of 'after care' products is required.

Problem	Cause	Possible solution
Client experiences irritation during processing.	Client may be sensitive to active ingredient in the formula.	Perform a 24-hour skin test and use a chemical preparation compatible with the client.
Hair is broken at the scalp level.	Too much tension was used when winding the hair onto the curlers. Curlers not secured correctly.	Do not wind with excessive tension. Securing mechanism on curler should not cut into the hair or be twisted against the curl.
If the desired result is a tight, springy curl, and this is not achieved.	A straight curler was used. Curl formation process was not complete. Curlers used were too large.	Use concave shaped curlers. Complete each stage of processing before proceeding to the next. Select correct size of curler for length and thickness of the hair.

CHAPTER 5
Hair Colouring

Hair colouring

Natural hair colour is an inherited characteristic due largely to the pigment granules found in the cortex. Human hair contains three different types of pigment:

1. melanin – a black or brown pigment
2. pheomelanin – a yellow or brownish-yellow pigment
3. trichosiderin – a red pigment.

The amount, distribution and type of pigment present varies from Afro-Caribbean people to Asian and European people, with melanin being the only one that is present in all races. It is also responsible for skin colour. Mixtures of the different pigments produce various shades of colour. The colour is actually produced by the pigments absorbing certain wavelengths or colours of light and reflecting others which become the visible colour. In this respect pigment is only one half of a pair that must exist in order for colour to be seen, with light being the other essential component. Pigment is the passive component which does not change, with light being the active partner, constantly changing. It acts on the same pigment in different ways under diverse conditions to produce different colours and tones. Thus hair appears to change colour when seen in bright sunlight, artificial light or candlelight for example.

Wet hair appears to be darker than dry hair as does hair that is oily. Dark brown hair usually has a reddish tinge which may not be apparent when it is wet. The hair of swimmers is usually lightened by the presence of chlorine in the swimming pools; this can have a very damaging effect on hair that has been relaxed, curly permed, lightened or tinted.

It is normal for hair to lose its pigmentation slowly with age. Some people grey as early as their late teens while others grey in their eighties. Although greying is associated with age, it is not to be taken as a sign of physical abnormality, the distribution of pigment in the hair and skin being unique to each individual.

Selecting a colour

A good hair-colourist must be able to advise which colours and tones will best suit a client, whilst at the same time allowing for the client's personal feelings. There are several basic points that must be considered when selecting artificial hair colour. It is essential to remember that colours affect the complexion both by reflection and by absorption, the difference created by light and dark hair being considerable. Lighter hair makes the skin look darker, while darker hair makes the skin look lighter. Where the skin is very light, the use of a black colour should be avoided as it can make the client look hard, sallow and older.

Where reddish skin tones are present it is likely that the hair also has hidden red tones. Auburn colours will accentuate and highlight these tones and bring them out to their full potential.

If a client has a rich caramel colour complexion, and the hair is dark brown, an application of medium brown would produce a complementary colour with auburn highlights.

When the skin colour is medium brown to fair, dark brown or auburn colours can be attractively worn, whereas if the skin colour is medium to dark brown, then auburn, black and coffee may be the most complementary colours to select.

Sometimes black people may be found to have naturally sandy hair which they may describe as mousy or dull. Such hair shades may be treated with a light brown to bring out the gold without a bleached look. Some clients may want their hair colour to be blond. Here it is worth remembering that if the client has dark skin, unnatural-looking blond colours, such as those typically favoured by Europeans, should be avoided. Caramel, butterscotch, honey and beige-blond can be just as stunning when worn with the complexion which complements them best.

Clients who have a ruddy complexion should be advised against red colours. Ash colours would prove more attractive for them but would not be suitable for those with a yellowish or brownish complexion.

Reference to the International Colour Chart System should be made when consulting a client about hair colour. However, it is worth remembering that colour charts and examples shown on packaging can often be misleading in their results. Ink on paper is not the same as chemical dye on hair, and no printed representation can accurately represent what the hair is going to look like upon completion. Allowing for this, such guides can be used as a basis to work from; the International Colour Chart System can be obtained from most hair colour manufacturers.

The principles of hair colouring theory are illustrated in Figure 5.1. There are two categories of colours; the primary colours being red, yellow and blue with violet, green and orange making up the secondary colours. Between yellow and orange is the colour character gold from which golden browns and golden blonds are derived. Between orange and red are the warm characters from which coppers and chestnuts are derived. Auburn shades, which are also said to be warm in character, are derived from red.

Figure 5.1
The colour circle

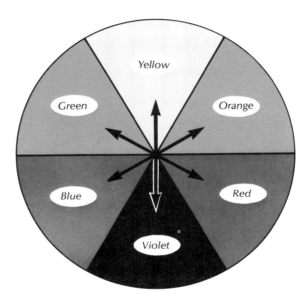

Between violet and blue is a colour character called cendre, between blue and green lies the colour ash; finally we come to green itself which is called matt. These are all usually referred to as cold or cool colours.

At the centre of the colour circle is neutral which is where the natural range of colours are located. Any colour in the circle may be neutralised by the colour located directly opposite.

There are many factors which need to be considered when selecting a colour, such as the client's age, dominant clothes' colours, personality and occupation. An extreme awareness of colour and co-ordination needs to be developed by the hair colourist whilst at the same time allowing for originality. The following outlines a good basic guide that should be followed when selecting a hair colour for a client.

1. Examine the client's hair in a well-lit area that is as close as possible to natural light.
2. Carry out the examination on dry hair which is free of oil or other material which may mislead someone trying to determine the natural colour of the hair.
3. Use a colour chart which displays the manufacturer's full range of colours including their names and reference numbers.
4. Select a colour with the client's co-operation and approval, having taken into consideration the client's age, skin undertones, complexion, life-style, occupation, personality, etc.
5. To match the client's hair colour, examine the hair nearest to the scalp at the back of the head.
6. When examining the hair for colour, look through it rather than down onto it. To determine the depth of colour, including highlights, raise the hair by pushing it up with the hands placed against the scalp.
7. Be familiar with the products which are being used.
8. Read and understand the manufacturer's instructions thoroughly and follow them strictly.

Testing for colour choice

Before applying a tint, do a strand test to determine the following:

- whether the correct colour has been selected
- how long the colouring (dye) should be left on the hair
- whether the hair is likely to break or discolour due to previous treatments such as relaxing, perming, the application of metallic or compound dyes, etc. If the hair is in a damaged condition, it should be treated using a suitable conditioning programme and re-tested before applying a tint.

Method for a strand test

1. Mix together a small amount of equal parts of the tint selected and 20 volume peroxide.
2. Apply the mixture to the full length of a dry hair strand and allow it to remain on the hair until the desired shade has developed. It is important to record the amount of time taken to achieve this.
3. Remove the tint from the test strand of hair, dry thoroughly and examine in a well-lit area of the salon.

If the results of the strand test are satisfactory, application of the tint may go ahead. However, if the colour produced is different from that desired, select another colour and do another strand test. If the results of the strand test show discolouration, which might indicate the presence of a metallic dye, then corrective action must be taken before application of the tint.

Types of dyes

The diverse range of hair colouring products available today can be classified into the following main groups:

1. temporary hair colour
2. semi-permanent hair colour
3. permanent hair colour   a) adding colour (tinting)
                            b) removing colour (lightening/bleaching).

Temporary hair colour

Products of this type colour the hair by coating the cuticular surface with a coloured material. Since the colour only remains on the surface of the hair it may be easily removed by shampooing. Products of this type are available in the following forms:

Colour rinses are used after shampooing and are allowed to dry on the hair. They are available either as a liquid or powder concentrate for all shades of hair including grey.

Colour shampoos are basically a combination of colour rinse and shampoo. They add highlights to the hair and are available in all shades.

Hair crayons are sticks of colouring which are available in all shades and compounded with soaps or synthetic waxes. They are used to retouch newly grown hair between tintings.

Hair colour creams are used mainly for theatrical make-up. They have a cream base which makes them easier to apply than crayons and are available in most colours.

Coloured lacquers consist of dyes combined with plastic resin dissolved in alcohol. They are generally applied from an aerosol can, leaving a film of coloured resin on the hair when the solvent evaporates. Lacquers are used mainly for exotic effects to complement fashion styles and are only available in a limited range of colours.

Temporary colouring products may be used to bring out highlights in the hair, restore faded hair to its natural colour, neutralise the yellowish tinge in white or grey hair, tone down hair which is too light or add colour to the hair without changing the condition of it. The disadvantages of temporary hair colour are that the colour is short-lived and must be re-applied after each shampoo, the coating is thin and may not give an even coverage, and the colour sometimes has a tendency to rub off onto clothing and pillows.

As there are many temporary hair colouring products available and many manufacturers, it is essential to read the instructions supplied with each product and follow them strictly.

Semi-permanent hair colour

This type of hair colour requires no peroxide developer and has a mild penetrating action. Such hair colours may require a patch test; this can be determined from the manufacturer's instructions. Semi-permanent tints are designed to do the following:

- cover or partially blend grey hair without affecting its natural colour
- enhance the beauty of grey hair without changing its colour
- highlight and bring out the natural colour of the hair.

Semi-permanent hair colour normally consists of nitro dyes which give red and yellow colours, and anthraquinones which give blue colours. Mixtures of these three colours produce a wide range of different shades. The molecules are small enough to pass into the cortex where most of them combine with the hydrogen bonds in keratin. Penetration into the cortex is aided by making the dye mixture alkaline, which has the effect of opening the cuticle. With each wash water replaces some of the dye molecules in the bonds and the dye washes out. Generally, the dye lasts for approximately 6–8 shampoos. Some tints contain ammonium thioglycolate in a weak solution, application of which onto relaxed hair could soften the 'S' bonds and damage the hair.

Semi-permanent colouring products are not ideal for Afro or black hair. Not only can they upset the molecular structure of relaxed hair, but in terms of colour they can be somewhat unpredictable in results. This is particularly true in the brown range. Browns are usually made up of mixtures of red, yellow and blue molecules with the blue molecules being the largest. Depending upon the porosity of the hair, not enough may penetrate, resulting in a brown quite different from that which was originally sought. If the hair is very soft and porous, semi-permanent tints should be avoided because so many blue molecules will penetrate and the result may be a greenish tinge. It is important to remember that the hair becomes very soft and porous after treatments such as perming. In such cases the colours wash out and fade at different rates resulting in the hair colour changing with time.

Permanent hair colour

There are four main categories of permanent colouring dyes. The most commonly used on black or Afro type hair is the oxidising tint containing an aniline dye which is also known as a coal-tar tint. The four main categories are:

1. *Aniline derivative tints* are also known as penetrating tints, synthetic organic tints, peroxide tints, oxidation tints, para- and amino tints. Toners are also classified as penetrating tints.

2. *Pure vegetable tints* are obtained from the flowers or leaves of plants and have a very limited range of colours. In the past, indigo, camomile, sage and Egyptian henna were used for hair colouring. The main vegetable dyes in present use are camomile, walnut and henna.
   Camomile (or chamomile) is derived from the dried flowers of the camomile plant, the active ingredient being apigenin. Its molecules are too large to enter the cortex so they coat the shaft as a yellow deposit. Camomile is used as a rinse after shampooing or is added to hair brightening and lightening shampoos and is only suitable for use on blond or light brown hair.
   Walnut dyes are derived by crushing unripe walnut shells, and produce a brown colour. They are added to modern colour shampoos for use on dark brown hair to deepen the colour.
   Henna is obtained from the dried leaves of the Egyptian privet, the active ingredient being lawsone. The dye molecules penetrate the hair shaft and are oxidised slowly with the final shade being reached 1–2 hours after application. Henna is also used in modern colour shampoos. When used on brown or chestnut hair it produces auburn highlights; when used on black hair it produces red highlights which stand out particularly well in bright sunlight. Henna is the only vegetable dye that is still used professionally in the salon.

3. *Metallic or mineral dyes*, such as lead acetate or silver nitrate, gradually darken and so 'restore' hair colour. The application of such hair dyes

renders the hair unsuitable for other chemical treatments such as relaxers, curly perms, tints, lightening, etc.

4. *Compound dyes* are a combination of vegetable dyes and metallic salts which actually fix the colour. Compound henna, for example, consists of a mixture of henna and silver, copper or lead salts. The use of such dyes also renders the hair unsuitable for other hair treatments as described above.

The greater part of hair colouring is done with the use of oxidation dyes containing an aniline derivative. The earliest oxidation dyes were para-phenylenediamine and para-toluene diamine, black and brown respectively. A much wider range of colours is now available with the addition of new chemicals.

The dyes are prepared as liquids or creams which have to be mixed with hydrogen peroxide immediately before use. When applied to the hair the small colourless dye molecules enter the cortex with the hydrogen peroxide where the process of oxidation takes place to produce large coloured molecules which become trapped in the hair shaft, thus producing a permanent colour change (see Figure 5.2). To maintain the new colour it is essential to dye the new growth of hair.

Figure 5.2
The action of
oxidation dyes

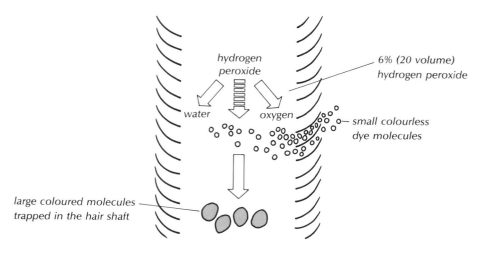

Tinting the hair with aniline derivatives does not interfere with other hair treatments. However, it is a wise precaution to consult the manufacturer's instructions before proceeding with such treatments.

There are two types of penetrating tints which are referred to as one-step and two-step tints. Lightening of the hair is not included as a separate step when using the one-step tint. The advantages of using the one-step method are as follows:

- It eliminates the need to pre-shampoo, pre-soften, or pre-lighten.
- It allows the client's hair to be coloured lighter or darker than the natural colour.
- It allows grey or white hair to match the client's natural colour.
- It can tone down or blend in streaks and off shades; and enliven discoloured and faded ends.

For general colouring purposes 6% (20 volume) hydrogen peroxide is used whereas 3% (10 volume) is used with cold colours such as ash, matt and cendre, for neutralising, and for pastel shades on bleached hair and more high fashion results. Where colour lift is required 9% (30 volume) peroxide should be used.

For reasons already described it is essential to be able to identify hair that has been treated with a metallic or compound dye. Such hair tends to be harsh and brittle to the touch and usually fades into peculiar or unnatural shades. Silver dyes have a greenish cast, lead dyes have a purple colour and dyes containing copper turn red. Dyes of this kind are usually used by people who have greying hair.

Highlighting shampoo tints

Highlighting shampoo tints are products that combine aniline derivative dyes, hydrogen peroxide and a neutral shampoo base. They are used to produce only a slight change in the natural hair colour. Some highlighting shampoos only contain hydrogen peroxide and a shampoo base. By omitting the tint and including only the lightener, colour is slightly removed from the hair. Where an aniline derivative is present a skin test is required before application can proceed.

The standard application technique for both the highlighting and lightening shampoo is:

1. Gown the client in the correct manner.
2. Position the client at a backwash basin.

3. Apply the mixture evenly on the client's hair and allow it to remain for the time recommended in the manufacturer's instructions.
4. Rinse thoroughly with warm water.
5. Dry and style as required.

Pre-softening

Pre-softening is recommended for grey or white hair that is resistant and will not readily absorb dye. Pre-softening such hair makes it more receptive to the dye and ensures more effective coverage of grey and white hair. There is no noticeable change in the colour of hair during this process.

Pre-softening mixture, which may be either 10 volume peroxide or an oil bleach, should be applied in the same way as hair colour, using a tint brush or bottle applicator. No pre-shampooing is necessary. The mixture should be applied to the entire length of the hair, and to the new growth only when performing a re-growth application. Allow it to remain for 20 minutes and then shampoo gently to remove the softener. Dry the hair with a towel and apply the tint mixture.

Bleaching

To lighten hair colour, the melanin must be oxidised by using a bleaching product with as little oxidation as possible of the disulphide linkages to minimise hair damage. Most bleaching products are added to hydrogen peroxide, usually in a 6% solution which releases 20 volumes of oxygen from each volume of itself. Bleaching products also contain an alkaline substance, usually ammonium hydroxide solution or solid ammonium carbonate to act as catalysts to release oxygen faster. The alkaline substance also acts as a wetting agent and aids entry of the bleach into the cortex.

Hydrogen peroxide is available in liquid or cream form with a wide variety of additives. In cream forms it is often called a 'developer'. Lanolin derivatives are used for thickening, clouding and emulsifying the clear hydrogen solution for appearance. In the cream form, it can provide more protection for hair and scalp, and it is easier to apply.

Types of bleach

Bleaches are classified as oil, cream, powder or paste bleaches.

Oil bleaches

Some oil bleaches consist of a sulphonated oil and hydrogen peroxide but most now contain an alkali, usually ammonium hydroxide, a wetting and gelling agent, to replace the role of the sulphonated oil as a thickener. Some oil bleaches also contain dye, adding temporary colour and highlights to the hair as they lighten. The colours contained are commercially made and may be used without a patch test. They remove pigment and add colour at the same time and may be classified in the following way:

- gold – lightens and adds gold highlights
- red – lightens and adds red highlights
- drab – lightens and adds ash highlights and reduces red and gold tones.

Ordinary oil bleaches may be used to pre-soften or lighten the hair prior to the application of the dye.

Cream bleaches

These are the most popular types of bleaches as they are easy to apply, do not run, drip or dry out. They contain conditioning, bluing and thickening agents which provide the following benefits:

- The conditioning agent provides some degree of protection to the hair.
- The bluing agent helps to reduce red and gold tones.
- The thickener provides more control when applying the bleach.

Powder or paste bleaches

As the term implies such products consist of a mixture of powders, magnesium carbonate and ammonium carbonate, which acts as the catalyst. For general bleaching purposes, the two powders have to be mixed together, immediately prior to use, with 6% hydrogen peroxide. Powder bleaches mix into a smooth paste which does not run. It does, however, tend to dry out quickly and, since the hydrogen peroxide is not diluted by any other liquid, such bleaches are more likely to cause damage to the hair than the other types of bleaches described.

Choice of bleach

When choosing what type of bleach to use always refer to the manufacturer's instructions. However, the following should be used as a general guide:

- Use an ordinary oil bleach to lighten the hair without adding any colour, as modern oil bleaches are more efficient and also leave the hair in good condition.
- Use an oil bleach (drab tones) to avoid red and gold highlights in the hair.
- Choose an oil lightener (red and gold shades) to obtain red or gold highlights. Choose a powder lightener for tipping, streaking or frosting extremely resistant hair (see Figure 5.3).

Figure 5.3
Highlighting
technique using cap

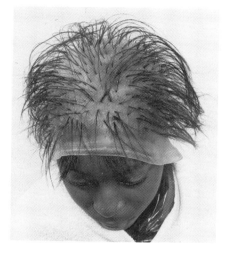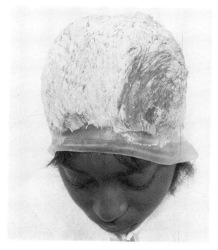

Application technique for virgin hair

1. Gown the client in the correct manner and examine the hair and scalp for any scratches, sores, cuts, bruises, etc. If any are present, do not proceed.
2. Divide the hair into four sections as previously described.
3. Apply a protective base all around the hairline.
4. Prepare the hair lightening mixture for immediate use and apply to the hair, starting at the crown. Apply the lightener 2.5 cm (1 in.) away from the scalp, covering each hair thoroughly except for the ends which are usually more porous and more likely to suffer from over-processing.
5. Continue the above application technique until all four sections have been covered. Speed is essential in order to obtain a uniform shade without streaks.
6. Take frequent strand tests and when the lightening process has reached the halfway point, apply the lightener to the one-inch hair next to the scalp. Begin this process at the same point as the initial application and when complete apply the lightener to the hair ends.

7. Pile the hair loosely on top of the head and continue the process until the desired shade is achieved. During processing it is important to keep the hair moist with the lightener.

8. When the desired shade has been achieved, rinse the lightener from the hair using warm water and then shampoo gently with mild shampoo.

9. Dry the hair with a towel or cool dryer and examine the hair for any signs of breakage. If everything is satisfactory application of the toner or colour may proceed. If the shade obtained is not the one desired, application of the lightener may be repeated after 24 hours. This is to allow for the client's comfort and also for safety.

Re-touch procedure

In time it becomes necessary to treat the new growth of hair to match the rest of the lightened hair. In re-touching, the lightener is applied to the new growth only, except when a lighter colour is required or if the colour becomes dull from several applications.

The application procedure is the same for a re-touch as for virgin hair, the only difference being that the lightening mixture is applied only to the new growth of hair. Care should be exercised to ensure that the lightener does not come into contact with previously lightened hair.

Problems in hair lightening

The natural colour of hair is determined by its pigment which may be black or brown, or red or yellow, or a combination of these pigments. Black and brown pigment change to a lighter shade in a few minutes after application of the lightener.

Red and yellow pigments are present in diffused form in the cortex of the hair. The diffused red pigment creates problems in lightening and tinting, making it difficult or impossible to lighten the pigment completely without causing great damage to the hair.

It is advisable not to guarantee that black or very dark hair can be lightened to a very pale blond shade (see Figure 1.9).

Whatever the reason for lightening the hair, it is important to choose the correct bleach and mixture for the degree of colour change desired. Use the colour chart and follow the manufacturer's instructions.

CHAPTER 6
Conditioners

Conditioners

Many hair treatments require the use of various chemicals which can affect the condition of the hair. Some chemicals, for example, remove moisture and oils from the hair, causing it to become dry and brittle. Such hair is referred to as damaged hair. Many people have hair that is naturally too dry, making it unmanageable and difficult to style; others have hair that is too greasy and which tends to lack body and looks limp. In all cases the hair can be treated with conditioners to restore the correct balance, making it look more attractive and healthy, easier to manage and suitable for various hair treatments.

Conditioners are available in either cream or liquid form, and may contain lanolin, cholesterol, moisturisers, enzymes, sulphonated oil, vegetable oils and protein, the formula varying according to the manufacturer. Most conditioners are applied to hair that has been shampooed and towel dried; precise details of application can be obtained from the instructions supplied with each product.

There are four main types of hair conditioners available and selection of the type to be used depends on the texture and condition of the hair and the results required.

Timed conditioners

These conditioners add natural oil and moisture to the hair without actually penetrating the hair shaft. When applied to the hair they are allowed to remain on for anything between 1–5 minutes before being rinsed out. The hair may then be styled with setting lotion as required. Timed conditioners usually have an acid pH.

Conditioners combined with styling lotions

Such hair conditioners act on the hair by coating each strand, increasing the diameter and thus giving the hair more body. They are available in several strengths to suit hair of different texture, condition and quality. Protein- or resin-based conditioners are incorporated into the setting lotion and applied as part of the hair setting process. A little water added during the hair setting procedure helps to facilitate setting by keeping the hair soft and manageable.

Protein penetrating conditioners

These conditioners, which utilise hydrolysed protein, pass through the cuticle and replace keratin that has been lost from the hair. They improve texture, equalise porosity throughout the length of the hair shaft and increase elasticity. The excess conditioners may be rinsed from the hair before setting in accordance with the manufacturer's instructions.

Neutralising conditioners

Strongly alkaline hair products can damage the hair and cause scalp irritation. Neutralising conditioners have an acid pH and are designed to neutralise such conditions. When applied to the hair they are allowed to remain on the hair for 1–5 minutes before being rinsed out.

CHAPTER 7
Afro Hair Cutting

Afro hair cutting

Regardless of ethnic origin or the type of hair an individual has, it is often cut for the same reasons, which may be one or a combination of the following:

- to shape the hair
- to thin the hair to remove excess bulk
- to reduce the length and remove split and damaged ends
- to produce the basic shape for a specific hairstyle.

The two main tools used for cutting are the scissors and the razor. The scissors can be used on wet or dry hair, but the razor can only be used on wet hair. Clippers may also be used on Afro hair.

Afro hair has its own special characteristics, as have other types of hair, which require special techniques for styling. The ability to create a hairstyle that will enhance the appearance of the client is quite obviously of primary importance. In this connection the stylist should be able to visualise how the client will look with the finished style. Knowing the correct shaping and styling techniques, and using common sense in their application are basic to the success of the hairstylist.

Figure 7.1
An off-the-face
hairstyle

Figure 7.2
Short version of the
'Afro' hairstyle

The main points

1. Comb the hair upward and slightly forward using an Afro comb or a wide-toothed comb, making the hair as long as possible. Start at the crown and continue until all the hair has been combed out from the scalp and distributed evenly around the head. To avoid sectioning, it is best to work around the head in a circular pattern.

2. Visualise the style and length of hair desired. Get the *shape* right first and then concentrate on the length. Begin by tapering the sides, cutting in the direction in which the hair will finally be combed.

3. Taper the back part of the hair to blend with the sides and trim the hair at the crown and the top of the head to the desired length.

4. For an off-the-face hairstyle, comb the hair up and backward, and for forward movement comb the hair up and forward. Blend the hair at the sides with the hair at the top, crown and back. Outline the hairstyle at the sides, around the ears and in the nape area, using either scissors or clippers. Finally, comb the hair lightly; hairspray may be applied to give the hair a natural, lustrous sheen.

5. The hairline may require attention to eliminate unevenness, i.e. 'widow's peak'; or to create a well defined hairline. This is called outlining and is achieved by skilful use of a razor or clippers.

Figure 7.3

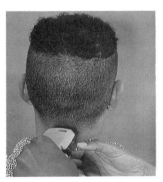  

Taper the nape *Blend in the side* *Finishing: cut the hair evenly at the crown and front of the head to the desired length*

Cutting with a razor

Using a razor to cut chemically relaxed or permed hair produces softer lines and gives a smoother finish. Using scissors on hair that has wave movement or soft (natural) curls can result in a chopped look. The use of a razor eliminates bulk without removing too much length.

Technique

1. Section the hair.

2. Starting at the back of the head and working towards the front, hold a section of hair in a comb and cut using the razor (see Figure 7.4). Next, cut at the sides, working from the front to the back. To prevent any injury to the client, hold the top of the ear down with one hand while working around it.

3. Complete the neckline by holding the hair down at the nape of the neck with the comb and cutting it with the razor.

Figure 7.4
Cutting the hair with a razor

 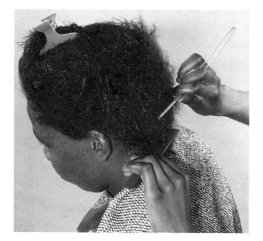

It is important to remember that if the razor is held at a 90 degree angle to the hair, it will only scrape it and not cut it. An angle between 20 and 30 degrees will cut deep into the hair while a 45 degree angle will not only cut the hair effectively but also safely. This is the correct angle at which the razor should be held.

Glossary

Acid	A compound which contains hydrogen atoms that are positively charged. These are called hydrogen ions and give acid its properties. An acid has a sour taste and turns litmus paper red.
Alkali	Also called a base, it is a combination of a metal with a hydroxide ion.
Allergy	Physical sensitivity to certain chemicals; skin rash.
Afro hair	Crinkly, curly hair usually characterised by its unruly nature.
Amino acid	An organic acid which combines its molecules into long chains to form proteins.
Amino tint	A synthetic organic chemical dye also called aniline dye.
Ammonium hydroxide	A chemical formed when ammonia is used to make ammonia thioglycolate.
Ammonium thioglycol(l)ate	A chemical used in cold wave lotion and curly perm (wet-look) hair preparations.
Aniline dye	Amino dye/coal tar tint.
Atom	The smallest particle of an element which still retains all of the properties of that element.
Base	Also called an alkali, it is a combination of a metal with a hydroxide ion.
Base (protective cream)	A petroleum-based protective cream used to protect the skin around the hairline and also used to protect the scalp with certain relaxers.
Bleaching	Removing or decreasing the natural colour from the hair; also called hair lightening.
Calcium	A vital element contained in bones and teeth. Milk is a source.
Calcium hydroxide	An active ingredient in hair relaxers.
Camomile or chamomile	A herb used for the purpose of brightening blond hair.
Capillaries	Tiny blood vessels which connect the arteries and veins.
Carbon	A basic substance which predominates in organic compounds.
Caucasoid	Refers to the racial (or ethnic) group of people who have the physical characteristics of the light-complexioned (white) peoples originating from Europe, North Africa, South-West Asia and the Indian subcontinent.

Cell	The smallest structural unit of an organism that is capable of independent functioning.
Colour rinses	Temporary colours, which are water based, used to colour the hair.
Colour shampoo	A semi-permanent colour; lasts through several shampoos.
Coloured lacquer	Temporary colour in a hair spray.
Compound dye	Type of dye used to darken the hair colour.
Conditioner	Any substance used to improve the existing hair condition.
Consultation	Meeting to discuss, identify and seek a practical solution to a problem with another person.
Cortex	The layer of hair underneath the cuticle.
Curl activator	An after-care product recommended for use on hair which has been chemically treated with a wet-look or curly perm.
Curl moisturiser	After-care product designed to prevent dryness and maintain moisture balance.
Curly perm	Hair which has been treated with ammonium thioglycolate to produce a cold wave on Afro hair.
Cuticle	Outer layer of the hair.
Cysteine	An amino acid containing sulphur which is easily oxidised to cystine.
Cystine	A crystalline amino acid containing two sulphur atoms found in proteins (as keratins).
Detergent	A cleansing substance made from chemical compounds, but not made like soap.
Disulphide bond	The polypeptide chains of hair molecules which are broken by the action of cold wave lotions.
Dye	Hair colour chemical; also called 'tint'.
Elasticity	The ability of hair to stretch without breaking.
Emigrate	To leave one country or region to settle in another. To move away from.
Enzyme	A substance which makes a chemical change in another substance without changing its own chemical make-up.
Ethnic	Pertaining to racial groupings based on common physical characteristics, such as skin colour, hair texture and bone structure.
Follicle	Skin containing the hair root.
Frosting	Sometimes called highlighting; technique of lightening pre-selected strands of hair throughout the entire head.
Genes	DNA units responsible for passing physical characteristics from one generation to the next.
Genetics	The biology of heredity, i.e. the study of hereditary transmission and variation.

Glycerine (hair coated by)	A common ingredient of hair preparations which are designed to make hair soft, and prevent dryness.
Granules	Small grains.
Hair	Threadlike growth covering the scalp.
Hair colouring	Lightening or darkening the hair by chemical means.
Hair colour creams	Cream-based hair colouring preparations used to cover new growth especially around the hairline.
Hair crayon	Hair colour in a 'stick' form used as a touch-up around the hairline to cover new growth.
Hair lightening	Bleaching the hair to a lighter colour. Removal of the natural pigment.
Hair relaxing	Reducing the natural curl in Afro type hair by chemical means. Texturising.
Henna	Vegetable compound dye used to colour hair.
Hydrogen	A gas which has no odour, taste or colour.
Hydrogen peroxide (neutraliser)	An oxidising agent used for bleaching, developing tints and neutralising permanent waves.
Immigrate	To settle permanently in another country.
Inflammation	Swelling, pain, skin fever; usually associated with irritation, bruising or abrasions.
Keratin	The basic substance which makes hair.
Lanthionine link	The sulphur atom of a sodium hydroxide relaxer takes on a different characteristic than that of a 'thio' relaxer. The cystine or sulphur atom remains independent and does not re-link or re-bond with its closest neighbouring bond. This new bond is called a lanthionine link.
Massage (hair)	Rubbing, stroking, manipulating hair or scalp during shampoo.
Medulla	The inner core of the hair structure.
Melanin	The substance in the cortex which gives natural hair its colour; decreases with age.
Metallic dye	Used to cover grey hair; usually applied daily it is 'built-up' on the surface of the hair.
Molecule	The smallest particle of an element that can exist alone and independent. See *atom*.
Mongoloid	Refers to the racial (or ethnic) group of people who have the physical characteristics of the peoples of the Far East, such as the Chinese and Japanese, and also including the Eskimos and the North American Indians.
Monitor	To observe development or progress closely.
Negroid	Refers to the racial (or ethnic) group of people who have the physical characteristics of the dark-skinned (black) peoples originating from Africa.
Neutraliser	Solution applied after hair relaxing or permanent waving.
Nitrogen	A gas found in living tissue and in the air.

No-lye (relaxer)	A term applied to relaxers which do not contain sodium hydroxide as an active ingredient.
Oxygen	Tasteless, colourless and odourless gas.
pH (potential of hydrogen)	The pH scale denotes the degree of acid or alkaline present.
Papilla	A small part at the bottom of the hair follicle.
Patch test	A skin test to determine sensitivity to specific chemicals.
Permanent wave	Changing the structure of the hair by a chemical process and re-shaping by the use of perm rods.
Peroxide	Means the same as hydrogen peroxide.
Pick	Name given to Afro comb.
Pigment	Organic colouring.
Polypeptide	Strings of amino acids joined together by peptide bonds, the prefix 'poly' meaning 'many'.
Porosity	Sponge-like; relates to the ability of the hair to absorb liquids.
Potassium hydroxide	Active ingredient in hair relaxer.
Precautions	Courses of action taken to ensure safety and avoid damage or accidents.
Pressing	The use of thermal combs to straighten Afro type hair.
Record card	Card on which details of client services and results for future reference are written.
Re-touch	New growth; also called 're-growth'.
Salt	The result of mixing a base with an acid; also sodium chloride, known as table salt.
Scalp	The skin covering the head, from which hair grows.
Sebum	Oil secreted from the sebaceous gland.
Sectioning hair	Dividing the hair into parts prior to application technique.
Shampoo	Technique for cleansing the hair. Preparation used to cleanse the hair.
Sodium bromate (neutraliser)	Used to neutralise the hair following a curly perm or wet-look.
Sodium hydroxide (lye)	Active ingredient in hair relaxer.
Spiral	Winding as in the threads of a screw.
Strand test	Test to determine the length of time a specific chemical may safely remain in contact with the hair in order to achieve desired results.
Streaking	Lightening wide or broad section of hair, attractively placed around the head and face.
Sulphur	A pale yellow, non-metallic element used in the preparation of hair preparations and rubber hardening.
Symptoms	Outward signs of inner disturbance or illness.
Test curl	In permanent wave it is called 'S' curl; it lets the stylist know that the process is complete.

Texture	Usually identified by the sense of 'touch' i.e. smooth, soft, hard, coarse, etc.
Thermal hair straightening	Also called 'pressing'.
Thioglycol(l)ate	An ingredient in permanent wave lotion.
Tipping	Hair colouring whereby the darkening or lightening is confined to the very ends of small strands of hair.
Transparent	Something one can see through.
Violet-ray	The pinkish violet discharge from high frequency equipment used for scalp stimulation.
Virgin hair	Hair which has not been chemically treated and is in its natural state.
Widow's peak	Irregular hair growth which comes to a point at the centre of the forehead.

Index